STARK COUNTY DISTRICT LIBRARY

DISCARD

INDEX

Africanized honeybees 6, 7
aggression 8, 12
allergies 8
Asian giant hornets 10, 11
blood 14, 16, 18, 21
boils 12
Chagas disease 18
chemicals 8, 10
colonies 13
conenose bugs 18
fire ants 12. 13
killer bees 6, 7
kissing bugs 18, 19
malaria 14

mosquitos 14, 15
parasites 4, 14, 16, 18
poop 18
sleeping sickness 16, 17
speed 7, 10
stingers 8, 10
swarms 4, 9, 12
territory 8
tsetse flies 4, 16
venom 4, 6, 8, 10, 12, 21
viruses 4, 14, 15
wasps 8, 9, 10
yellow jackets 9

FOR MORE INFORMATION

Books

Goldish, Meish. *Red Imported Fire Ants: Attacking Everything*. New York, NY: Bearport Publishing, 2016.

Pallotta, Jerry. *Hornet vs. Wasp*. New York, NY: Scholastic, 2013.

Pearson, Scott. *Africanized Honeybees*. North Mankato, MN: Black Rabbit Books, 2017.

Websites

Insects
www.dkfindout.com/uk/animals-and-nature/insects
Discover much more about insects here, including how different insects harm and help people.

Mosquito
kids.nationalgeographic.com/animals/mosquito
Find out more about mosquitoes, the little insects known for spreading many of world's deadliest diseases.

Tsetse Fly Facts
www.softschools.com/facts/animals/tsetse_fly_facts/1250/
Learn more facts about tsetse flies.

Publisher's note to educators and parents: Our editors have carefully reviewed these websites to enscied.ucar.edu/webweatherwever, and we cannot guarantee that a site's future contents will continue to meet our high standards of quality and educational value. Be advised that students should be closely supervised whenever they access the internet.

GLOSSARY

aggressive: showing a readiness to attack
allergic: to have a sensitivity to usually harmless things in the surroundings, such as dust, pollen, or insect stings
chemical: matter that can be mixed with other matter to cause changes
fever: a body temperature higher than normal
mate: to come together to make babies
mound: a small hill or pile of dirt or stones
organ: a part inside an animal's body
parasite: a living thing that lives in, on, or with another living thing and often harms it
swarm: to move somewhere in large numbers
territory: an area of land that an animal considers to be its own and will fight to defend
tissue: matter that forms the parts of living things
venomous: able to produce a liquid called venom that is harmful to other animals

THE INSECTS IN THIS BOOK ARE LETHAL, BUT THEY AREN'T THE ONLY ONES THAT CAN HARM YOU. BE CAREFUL OUT THERE!

DRIVER ANTS

OTHER DANGEROUS INSECTS

INSECT	HOW IT'S HARMFUL
PUSS CATERPILLAR	VENOM MAKES SKIN FEEL LIKE IT'S BURNING
BEDBUG	HIDES IN BEDDING AND SUCKS BLOOD
DRIVER ANT	STRONG MOUTHPARTS DELIVER A PAINFUL BITE
HUMAN BOTFLY	LARVAE GROW UNDER THE SKIN OF HUMANS AND ANIMALS
ROVE BEETLE	RELEASES A TOXIN THAT BURNS SKIN AND EYES IF CRUSHED
FLEA	CAN CARRY AND SPREAD DISEASES

KILLER INSECTS

There are more than 5 million species, or kinds, of insects on Earth. And new species are being discovered all the time!

Luckily, only a small number of insect species are dangerous to humans. Some even help us—but other bugs should watch out! Some insects hunt and eat pests that harm crops. Other insects lay their eggs on harmful insects. When baby insects come out of the eggs, they enter the harmful insects and eat them alive!

KISSING BUGS HAVE MANY HIDING PLACES. THEY'VE BEEN FOUND UNDER CEMENT, BENEATH PORCHES, AND IN HOLES IN WALLS.

THE KISS OF DEATH

Conenose bugs are insects better known as "kissing bugs." They usually live in the southern United States, Mexico, Central America, and South America. Their nickname is cute—but what they do isn't sweet at all.

THE FORCE OF NATURE

IT'S BELIEVED THAT UP TO 8 MILLION PEOPLE IN THE AMERICAS HAVE CHAGAS DISEASE. FOR AROUND 30 PERCENT OF THOSE INFECTED, THE DISEASE WILL CAUSE PROBLEMS THAT MAY PUT THEIR LIFE IN DANGER.

Kissing bugs will often bite you near your mouth or eyes. These bugs drink your blood and then poop on or near the bite! The poop of a kissing bug may carry a parasite that can give you a sickness called Chagas disease. This disease can cause deadly heart problems.

IN THE PAST, AROUND 10,000 CASES OF SLEEPING SICKNESS WERE RECORDED EVERY YEAR, BUT THAT NUMBER HAS DROPPED A LOT. IN 2015, FEWER THAN 3,000 CASES WERE REPORTED.

TERRIBLE TSETSE FLIES

If you have food or drinks outside, flies sometimes come over and bother you. This might make you a little mad, but these flies aren't dangerous. Africa's tsetse flies, however, are lethal insects that suck your blood. They spread a parasite that causes a deadly disease called sleeping sickness.

When you have sleeping sickness, you get a high fever and your head starts to hurt. Then, you aren't able to think or move well. Finally, you become sleepy. Death often follows.

THE FORCE OF NATURE

THERE ARE TWO FORMS OF SLEEPING SICKNESS. THE STRONGER FORM OF SLEEPING SICKNESS CAUSES DEATH WITHIN MONTHS. SOMEONE WITH THE WEAKER FORM OF SLEEPING SICKNESS MAY TAKE SEVERAL YEARS TO DIE.

MOSQUITOES SPREAD MANY OTHER DISEASES, OR ILLNESSES, INCLUDING WEST NILE VIRUS, ZIKA VIRUS, AND DENGUE.

MENACING MOSQUITOS

Quick—name the deadliest animal on Earth. Would you be surprised to learn it's the mosquito? Mosquitos are small flies that feed on the blood of some animals and can spread illness. They cause about 725,000 deaths a year because of the viruses and parasites they pass on to people.

Mosquitos can pass along malaria, one of the world's deadliest illnesses. If you have malaria, you'll have a high **fever** and chills. In bad cases, your brain swells, and your **organs** fail. These effects can cause death.

THE FORCE OF NATURE

ARE YOU SCARED OF SHARKS? SHARKS KILL FEWER THAN 10 PEOPLE A YEAR. MALARIA, HOWEVER, KILLS OVER 400,000 PEOPLE EVERY YEAR AND MAKES ANOTHER 200 MILLION PEOPLE VERY SICK.

FIERCE FIRE ANTS

You may have seen tiny ants in your kitchen looking for food. These ants are bothersome, but they won't hurt you. But fire ants can cause great pain!

THE FORCE OF NATURE

WITHIN 1 TO 2 DAYS, A FIRE ANT'S STING BECOMES A FIRM BOIL, OR SWOLLEN BUMP UNDER YOUR SKIN. THESE BOILS CAN BECOME INFECTED, OR FILLED WITH GERMS.

Fire ants aren't big, but they're aggressive and have a powerful sting. At the first sign of danger, such as the sound of your footsteps, they swarm out of their **mounds** and attack. Fire ants grab your skin with their mouthparts, stinging you again and again. Their strong venom causes a burning feeling.

EACH YEAR, AROUND 30 TO 50 PEOPLE IN JAPAN DIE FROM BEING STUNG BY ASIAN GIANT HORNETS.

ASIAN GIANT HORNETS

Hornets are a type of wasp. The world's largest hornet is the Asian giant hornet. They can grow to around 2 inches (5 cm) long and chase you at 25 miles (40 km) per hour! Could you outrun one?

Asian giant hornets have more pain-causing chemicals in their venom than any other stinging insect. Their venom also has a chemical that can break down your body's **tissue**. And, like all wasps, these hornets can sting again and again!

THE FORCE OF NATURE

THE ASIAN GIANT HORNET HAS A STINGER THAT'S 1/4 INCH (0.64 CM) LONG!

WICKED WASPS

Wasps and bees seem very similar. They both belong to the same insect order, or group of animals. Both insects also sting their enemies. But bees can only sting an enemy once. After a bee stings you, it dies because its stinger pulls out its insides. However, wasps can sting you many times!

Wasps are famous for being **aggressive**. If wasps think you're a danger to their **territory** or young, they'll chase you for a long distance. They also give off a **chemical** that brings others to join the attack!

THE FORCE OF NATURE

IF YOU'RE **ALLERGIC** TO WASP VENOM, A SINGLE STING CAN BE DEADLY. A 2018 STUDY FOUND THAT THE STINGS OF WASPS, HORNETS, AND BEES KILL AROUND 60 PEOPLE EVERY YEAR.

AFRICANIZED HONEYBEES CAN MOVE FAST. THESE KILLER BEES FLY AROUND 15 MILES (24 KM) PER HOUR!

KILLER BEES

Africanized honeybees have a terrifying nickname: killer bees. These bees are the result of **mating** African honeybees with wild European honeybees. Africanized honeybees aren't bigger than European honeybees and their venom isn't stronger. So, what makes them "killers"?

If European honeybees think their hive is in danger, a few may fight to keep the hive safe. But, if killer bees think their hive is under attack, a huge swarm may fight back. Hundreds of bee stings can kill a human!

THE FORCE OF NATURE

AFRICANIZED HONEYBEES WILL CHASE AN ENEMY AS FAR AS 1/4 MILE (0.4 KM). IF YOU JUMP INTO WATER TO ESCAPE, THEY'LL WAIT FOR YOU TO COME UP FOR AIR!

LETHAL INSECTS DON'T ALWAYS LOOK SCARY, BUT SOME OF THEM ARE AMONG THE DEADLIEST CREATURES ON EARTH.

BRUTAL BUGS

Insects are small, often winged, animals with six legs and three main body parts. They're all around us. Some are beautiful, like butterflies. Others are bothersome, like flies. Some are even lethal, or deadly!

TSETSE FLY

Lethal insects can cause pain, suffering—and even death. Some use **venomous** stings to cause harm. Others carry illness-causing viruses or **parasites** they pass on to people. Some lethal insects only bother you if you bother them. Others attack in **swarms**. Watch out for these dangerous insects!

CONTENTS

Brutal Bugs 4
Killer Bees 6
Wicked Wasps 8
Asian Giant Hornets 10
Fierce Fire Ants 12
Menacing Mosquitos 14
Terrible Tsetse Flies 16
The Kiss of Death 18
Killer Insects 20
Glossary 22
For More Information 23
Index 24

Words in the glossary appear in **bold** type
the first time they are used in the text.

Please visit our website, www.garethstevens.com. For a free color catalog of all our high-quality books, call toll free 1-800-542-2595 or fax 1-877-542-2596.

LLibrary of Congress Cataloging-in-Publication Data

Names: Levy, Janey, author.
Title: Lethal insects / Janey Levy.
Description: New York : Gareth Stevens Publishing, [2020] | Series: Mother nature is trying to kill me! | Includes index.
Identifiers: LCCN 2018049570| ISBN 9781538239704 (paperback) | ISBN 9781538239728 (library bound) | ISBN 9781538239711 (6 pack)
Subjects: LCSH: Insects–Juvenile literature. | Insect pests–Juvenile literature. | Poisonous arthropoda–Juvenile literature.
Classification: LCC QL467.2 .L47 2020 | DDC 595.7–dc23
LC record available at https://lccn.loc.gov/2018049570

First Edition

Published in 2020 by
Gareth Stevens Publishing
111 East 14th Street, Suite 349
New York, NY 10003

Copyright © 2020 Gareth Stevens Publishing

Designer: Sarah Liddell
Editor: Monika Davies

Photo credits: Cover, p. 1 Glass and Nature/Shutterstock.com; background used throughout Oat Photographer ThaiLand/Shutterstock.com; pp. 4, 17 Patrick Robert - Corbis/Contributor/Sygma/Getty Images; p. 5 Arcaion/Shutterstock.com; p. 7 Hic et nunc/Wikimedia Commons; p. 9 Sean McVey/Shutterstock.com; p. 11 (main) File Upload Bot (Magnus Manske)/Wikimedia Commons; p. 11 (inset) NUMBER7isBEST/Wikimedia Commons; p. 13 MR. AUKID PHUMSIRICHAT/Shutterstock.com; p. 15 (main) Digital Images Studio/Shutterstock.com; p. 15 (inset) Kateryna Kon/Shutterstock.com; p. 19 Ava Peattie/Shutterstock.com; p. 20 Mark Schwettmann/Shutterstock.com; p. 21 Bartolucci/Wikimedia Commons.

All rights reserved. No part of this book may be reproduced in any form without permission in writing from the publisher, except by a reviewer.

Printed in the United States of America

CPSIA compliance information: Batch #CS19GS: For further information contact Gareth Stevens, New York, New York at 1-800-542-2595.

MOTHER NATURE IS TRYING TO KILL ME!

LETHAL INSECTS

BY JANEY LEVY

Gareth Stevens
PUBLISHING